Antimatter Propulsion

Case Study of Antimatter Propelled Interstellar Flights to Nearby Stars

Dr Ugur GUVEN

Guven Publications

Copyright

Antimatter Propulsion: Case Study of Antimatter Propelled Interstellar Flight to Nearby Stars

ACKNOWLEDGEMENT

The author thanks his students who have been instrumental in supporting this research and this book.

ABSTRACT

One of the few ways to do quick interstellar rendezvous missions is with a matter-antimatter annihilation propulsion system. This book discusses the general mission requirements and system technologies that would be required to implement an antimatter propulsion system where a magnetic nozzle (superconducting magnet) is used to direct charged particles (from the annihilation of protons and antiprotons) to produce thrust. Scaling equations for the various system technologies are used where, for example, system mass is a function of propulsion system power, and so on. Einstein's laws are used energy mass relation and relativistic effects. For the vehicle undergoing 0.1c velocity we will see the mass dilation, time dilation and length contraction at different distances on its way to different stars.

This book will portray a case of antimatter propulsion powered interstellar flight to Alpha Centauri as well as to Barnard's Star and to Wolf 359. Flying to these stars is truly an interstellar mission. The book is ideal for undergraduate engineering students and for Master level students interested in interstellar travel.

Table of Contents

1. <u>INTRODUCTION</u>

One of mankind's earliest dreams has been to visit the outer reaches of sky. Over the last 40 years we have visited most of the planets in our solar system, reaching out far beyond the orbit of Pluto with our unmanned spacecraft. And yet this distance, which strains the limits of our technology, represents a tiny step towards the light-years that must be traversed to travel to the nearest stars. For example, even though the Voyager spacecraft is one of the fastest vehicles ever built, traveling at 17 km/s or 3.6 AU/year, it would still require almost 74,OOO years for it to travel the distance to our nearest star. Thus, travel to the stars is not impossible; however, represent very hard work by a civilization simply because of the size and scale of any technology designed to accelerate a vehicle to speeds of a fraction of speed of light.

While antimatter has been a rich source of inspiration for writers of science fiction, it has also grabbed considerable attention in the astronautical community, where many studies have considered antimatter as a spacecraft fuel. The task of producing and safely storing antimatter in macroscopic quantities may prove to be unfeasible or prohibitively difficult and will undoubtedly be a very futuristic technology if it ever comes to fruition. On the other hand, the incomparable energy storage per unit mass of this potential fuel motivates a very long-term perspective when considering feasibility. Matter and Antimatter annihilation produces pure energy, the average energy produced in annihilation process is billion times greater than thermal energy, the concept of pellet was proposed by Cassenti in 1997. These pellets were 2cm in diameter and made use of transient magnetic fields to contain plasma of deuterium and tritium. For high specific impulse magnetic nozzles can direct the exhaust plasma.
Five and ten percent fusion will yield 15O,OOO seconds and 2OO,OOO seconds respectively, and fusion systems with smaller inert pellet masses and energetic fusion reaction products can give specific impulse of more than 1,OOOO,OOOseconds.

2. **THE VISION-MISSION**

For this project to reach to Proxima Centauri and other stars within a 10 light year distance we have taken a cruise velocity of O.1c.Our mission to reach Proxima Centauri and other stars include visions like fast interstellar velocity around O.1c based on fusion process. During the mission we will see the change in velocity at different distances both in Oort cloud where Sun's gravitational influence can be felt and the interstellar space in which the space craft leaves the Oort cloud and continues its journey to the star. We will see how this fraction of speed of light velocity will affect the aircraft at different distances and what measures should be taken during designing.

3. <u>ANTIMATTER ROCKET SYSTEMS ANALYSIS</u>

This project discusses the mission requirements of a beamed core process in which a Magnetic nozzle is used to direct charged pions which are formed as a result of proton and antiproton annihilation. These matter and antimatter annihilation will result in high energy photons gamma rays coming out as a process. Therefore, radiation shield must also be used.

The proton and antiproton annihilation will form charged pions, thus magnetic field has to setup to direct these charged pions because the formation of charged pions will be random and a magnetic nozzle will help him in directing the charged pion to produce the thrust required.

4. MATTER ANTIMATTER ANNIHILIATION

Antimatter is the exact replica of matter particles in terms of mass, size, everything except the charge thus when the antimatter comes in contact with its matter counterpart it annihilates to give back pure energy the annihilation could be of an electron-positron (counter part of electron) or proton and an antiproton. Electrons come under leptons, they are fundamental particles and hence its annihilation with positron will result in a complete annihilation process to give energy but a proton is a baryon and it comprises of two up quarks and one down part. So its annihilation with an antiproton will result in the formation of charged pions which can then be used by directing away from the nozzle to produce thrust.

The conversion of mass of both matter and antimatter particle to give energy follows Einstein's law $\boxed{E = mc^2}$.

$$M + AM \rightarrow ENERGY$$

This energy is in the form of photons. Since the mass of both matter and antimatter particle are same the total energy produced as by Einstein law will be equal to $2\,\boxed{mc^2}$

5. <u>ELECTRON–POSITRON ANNIHILATION</u>

When a low energy electron annihilates with a low energy positron it can only produce two or more gamma ray photons because the electron positron do not carry enough mass energy to produce heavier particles.

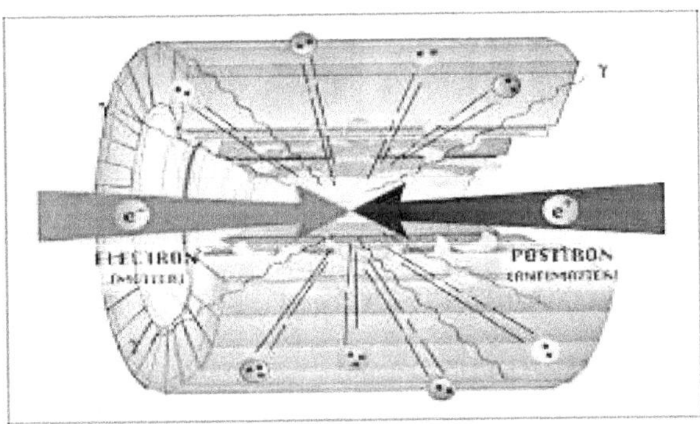

Fig 5.2: Virtual pair production

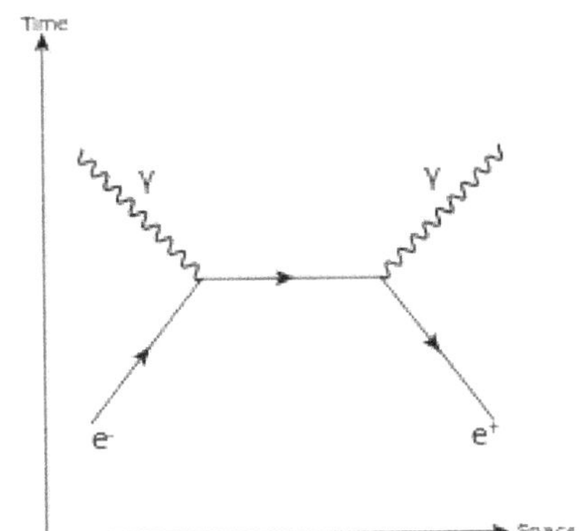

Fig 5.1: Electron positron annihilation Feynman diagram

5.1. The physics of anti-proton–proton annihilation rocket propulsion

Unlike electron proton is not a fundamental particle it can further be subdivided into two up quarks and one down quark. When a proton encounters its antiparticle the reaction is not simple as electron positron annihilation. The constituent valence quarks may annihilate with an anti-quark will the remaining quarks and anti-quarks will undergo rearrangements into number of mesons which will fly away from the annihilation point.

These newly created mesons are unstable and will result in decay in a series of reaction ultimately producing gamma rays. The energy released in reaction is substantial as the mass of three pions is much less than the mass of the proton and the antiproton.

6. <u>STORING ANTIPROTONS</u>

Protons and antiprotons are quite massive compared to the electrons and positrons, and the energy required to make them is also greater. As long as we have enough energy we can create antiprotons. But, controlling them once you have made them is a big challenge. First a beam of proton is fired into a block of metal. About once in every 250,000 collisions, kinetic energy is converted into mass in the form of a new antiproton–proton pair. These antiprotons travel close to the speed of light, moving in every direction. Magnetic fields that had been able to focus positrons into stable orbits were unable to control the randomly moving antiprotons, which would fly sideways, smash into the walls and be annihilated. Therefore some way was needed to control them. Electrons are destroyed by their antiparticle, the positron, while the antiproton is at risk only from protons or neutrons. The random movement of the antiprotons was gradually smoothed out as their energy was transferred to the electrons.

The Dutch engineer, Simon Vander Meer, shared his insight that when particles go around a curve, it takes longer for them to traverse a semicircle than to send a signal at the speed of light across the diameter. CERN implemented the idea in a small machine called an Antiproton Accumulator, known as the AA. As its name suggests, this took antiprotons and cooled them into a well-behaved beam, gathering and storing them until there were enough tube useful. Electronic detectors on opposite sides of the ring monitored where the antiprotons in the beam were travelling as they passed. The signal went to a computer which calculated by how much the beams were diverging and how big a kick was needed to align them better, and then sent a signal at the speed of light to electrode son the far side of the ring. The clever insight was that it takes about 50 per cent longer for antiprotons to travel around a semi-circle than for a signal to take the short cut across the diameter, and that if the ring was large enough, this would leave enough time for the electronics to work out what needed to be done, and to send the instructions, which the receivers act on, before the antiprotons finally arrived round the bend. In a billionth of a second, a 'nanosecond', light travels one-third of a meter.

Every two seconds, protons burst out of the PS, smashed into target and produced antiprotons. These entered the Antiproton Accumulator one burst at a time, where they were cooled for two seconds until the next burst arrived. The AA was like two rings in one, connected by shutters that could be opened and closed. On the ring inside the shutters, bunches of cooled antiprotons would be circulating, while outside the shutters were the latest newly arrived antiprotons that were still in the process of being cooled. Just before the next burst was due, the shutters would open and the antiprotons in the outer ring, now cooler, switched to the inner ring. The shutters were then closed, the next burst entered, and the process repeated over Andover.

Once the antiprotons were in the inner stack, van der Meer's electronic messages flashed across the ring, cooling them even This is easy to remember as it is about the size of an adult's foot, so if you can estimate how many of your feet would fit into some distance, that tells you how many nanoseconds light will take for the journey more. It took just over a day to accumulate and cool a hundred billion antiprotons. Van der Meer's trick led to intense beams of high energy antiprotons that could be used in experiments.

Antiprotons have greater mass and are harder to control than positrons, but once under control, the antiprotons are much better at creating energy than any other particle. Through the annihilation of antiprotons and protons, conditions of the first moments of the Big Bang can be created. Antiprotons can be controlled but the process is very slow and requires a big sum of money.

7. **RELATIVISTIC SPACE FLIGHT DYNAMICS**

For relativistic physics it is important to note that among all the existing reference frames there is one preferred frame: the rest frame. This is the frame of the object under consideration in which it is at rest. Any other external observer having velocity v relative to this rest frame observes the properties of the object such as length, time, speed, and acceleration differently as the object itself. Since there may be an infinite number of observers and therefore many different views of the object properties, relativistic physics holds that only one has a proper view of the object: the object itself. Therefore, relativistic physics introduces the notion of "proper". In general, a "proper" measure of a quantity is that taken in the relevant instantaneous rest frame, therefore also called proper reference frame. So "proper" is everything an astronaut experiences in his rocket. This is why we will not dash such quantities rocket. This is why we will not dash such quantities throughout this article and those as observed from outside will be dashed. Adapting this notion, what is of relevance first is how the proper measures relate to the measures of external observers.

7.1. Proper Time

Proper time is the time that the person in the will experience based on its clock however an observer from earth due to the relativistic speed of the rocket will experience dilation in time. The relation dilated time to proper time is given as

$$d\tau = dt = \sqrt{\frac{1}{1-\frac{v^2}{c^2}}}\, dt' \qquad\qquad eqn\ 7.$$

Where $\sqrt{\dfrac{1}{1-\dfrac{v^2}{c^2}}} = lorrentz\ factor$ **eqn 7.**

7.2. Proper Acceleration

The proper acceleration based on Einstein view is that acceleration is an absolute concept: an astronaut does not experience rocket velocity in his rest frame, but he does so for acceleration. Letups assume that the astronaut experiences acceleration a. Then special relativity tells us that this is related to the acceleration a' as seen by an external observer.

$$\alpha = \gamma^3 a' \qquad\qquad \textbf{eqn 7.}$$

7.3. Proper Speed

Because acceleration is an absolute concept we are apt to define

$$d\sigma = \alpha(t)dt \qquad\qquad \textbf{eqn 7.}$$

$d\sigma$ is an increase in speed as measured in the instantaneous rest frame. We integrate to get

$$\sigma(\tau) = \int_0^\tau \alpha(t)dt \qquad\qquad \textbf{eqn 7.}$$

This equation tells us that σ is the integral of the acceleration as experienced in the proper reference frame and hence is the speed as experienced by an astronaut, who sees the outer world going by. Since this is the true meaning of proper, σ is a proper speed. In order to find the relation of this proper speed to the relative speed v we apply

$$\sigma = \int_0^\tau \gamma^2 a' \, dt' \qquad\qquad eqn\ 6$$

$$as\ a' = \frac{dv}{dt'} = c\frac{d\beta}{da'}\ we\ fin$$

$$\sigma = \int_0^v \frac{dv}{1 - v^2/c^2} = c\int_0^\beta \frac{d\xi}{1 - \xi^2} = c\tanh^{-1}\beta \qquad\qquad eqn\ 7.$$

$$\mathrm{Or}\quad \beta = \frac{v}{c} = \tanh$$

It is now shown that the proper speed is proper in more general sense. Let us consider a second rocket or any other object in space having the known speed u relative to the astronaut's R system. We want to know what its speed u' is as measured by O'. Special relativity tells us

$$u' = \frac{u + v}{1 + \dfrac{uc}{c^2}} = c\frac{\beta_u + \beta_v}{1 + \beta_u\beta_v} \qquad\qquad eqn\ 7.$$

The problem with this transformation equation is that it is not linear as in classical physics where the Galileo transformation $u' = u + v$ holds. In addition, the above limits u' to the range $0 \leq u \leq c$ *if* v starts out from below c. This can be seen immediately if one inserts even limiting velocities $u=c$. This is Einstein's famous law that nothing goes faster than the speed of light. It is exactly this non-linearity and limited range of values which causes problems when treating special relativity mathematically.

The proper speed goes to infinity if the externally observed speed goes to the speed of light. This is to say that from an astronaut's point of view there is no speed limit. He actually can travel much faster than the speed of light. But of course he cannot travel faster than infinitely fast. This is the reason why the observer also sees a speed limit: the speed of light. So the ultimate reason why nothing can ever go faster than the speed of light is that nothing can ever go faster than infinitely fast. Note that from this point of view photons always travel infinitely fast. They experience that any distance in the universe is zero: for them the universe is one solid block. Because their proper time is zero one would say, they do not even exist. But this would be wrong. They come into existence at one point in our universe, they transfer energy, momentum, angular momentum, and information to any other point in proper zero time thereby causally linking any two parts in our universe and at the instance their work is done they are gone. This is why causality is the basic conservation law and hence the cement of our universe, and not the speed of light. The speed of light c may vary throughout our universe, but the fact that the proper time at $v = c$ is always zero and cannot become negative — implying that no inverse causality is possible — is firm

We summarize by noting that the proper speed exhibits four important properties: it is proper, it transforms linearly, it takes on all real numbers and it turns over into the classical concept of velocity at low speeds. This implies that it is a natural extension of the classical speed into special relativity and is mathematically integral.

8. <u>RELATIVISTIC ROCKET</u>

A relativistic rocket is any spacecraft that is travelling at a velocity close enough to light speed for relativistic effects to become significant. What "significant" means is a matter of context, but generally speaking a velocity of at least 10% of the speed of light (0.1c) is required. The time dilation factor, mass factor, and Lorentz contraction factor are O.995 at 0.1c. At speeds that are fraction of speeds of light or more Einstein's laws describes the motion of the body.

Achieving relativistic velocities is difficult, requiring advanced forms of spacecraft propulsion that have not yet been adequately developed. Nuclear pulse propulsion could theoretically achieve O.1c using current known technologies, but would still require many engineering advances to achieve this. The relativistic gamma factor (γ) at 1O% of light velocity is O.995. Generally a rocket that achieves at least 10% of light velocity or more is a relativistic rocket. The time dilation factor of O.995 which occurs at 1O% of light velocity is too small to be of major significance. At O.1Oc velocity interstellar rocket is thus considered to be a non-relativistic rocket because its motion is very accurately described by Newtonian physics also.

Relativistic rockets are usually seen discussed in the context of interstellar travel, since most would require a great deal of space to accelerate up to those velocities. They are also found in some thought experiments such as the twin paradox.

8.1. Relativistic Rocket Equation

With the concept of proper speed at hand we start out to derive the relativistic rocket equation. We want to do this in its most general form. Future relativistic rocket space flight must tap the ultimate energy source of nuclear fuel fission or fusion (including matter–antimatter annihilation) whereby mass is converted into energy according to $E = mc2$. On the other hand, the two physically distinct rocket propulsion systems are mass propulsion and photon propulsion. We take both into account, and assume that upon combustion a portion £ of the fuel mass will be converted into energy with a certain efficiency η and that a portion of it expels the exhaust mass with velocity v_e, while the other portion (1–ʒ) is expelled as exhaust photons, and the rest is lost. Therefore, the overall energy scheme looks like In the rest frame R of the rocket momentum conservation holds. Taking the momentums of both exhaust components and that of the rocket into account we can write

$$(1 - \varepsilon)dm.\,v_e + (1 - \delta)\eta\epsilon dm.\,c + (m + dm)dv = \qquad \text{eqn 8.1}$$

Since m is the mass of the rocket we have to count dm negatively. From the above equation we find

$$dv = -v_{ex}\frac{dm}{m}, \quad dm < \qquad \text{eqn 8.2}$$

With $v = (1 -£)\,v_e + (1 -ʒ\,)\,\eta£c$ the effective exhaust speed. Note that all terms are undashed and are therefore terms measured in the proper reference frame including dv. Now, in classical physics the relation $dv = dv'$ holds; and hence the equation can be readily integrated to yield the classical rocket equation $v = v_{ex}\ln(m_i/m)$ where index i indicates initial values and for convenience we have assumed $v_i = 0$. But $dv=dv'$ is no longer valid for relativistic speeds. However, if we identify $dv = d\sigma$ we again can directly integrate to obtain

$$\sigma = v_{ex}\ln\frac{m_i}{m} \qquad \textit{relativistic rocket equation} \qquad \text{eqn 8.}$$

with the same assumption $\sigma i = O$. So the relativistic rocket equation in this form is to the utmost extend complementary to the classical rocket equation. In order to show that above equation is in accordance with today's more convenient form of the relativistic rocket equation we apply and the algebraic equation for the free variable x.

$$\tanh^{-1} x = \ln\sqrt{\frac{1+x}{1-x}} \qquad eqn\ 8.$$

$$\frac{m_i}{m} = \left(\frac{1+\beta}{1-\beta}\right)^{1/2\beta_{ex}} \qquad with\ \beta_{ex} = {v_{ex}}/{c} \qquad eqn\ 8.$$

From the above equations we can also derive the thrust F of the relativistic rocket in its rest frame

$$F = ma = m\frac{d\sigma}{d\tau} = -v_{ex}\frac{dm}{d\tau} = -\dot{m}v_{ex}\ ,\quad \dot{m} \qquad eqn\ 8.6$$
$$< 0 \quad relativistic\ rocket\ thrust$$

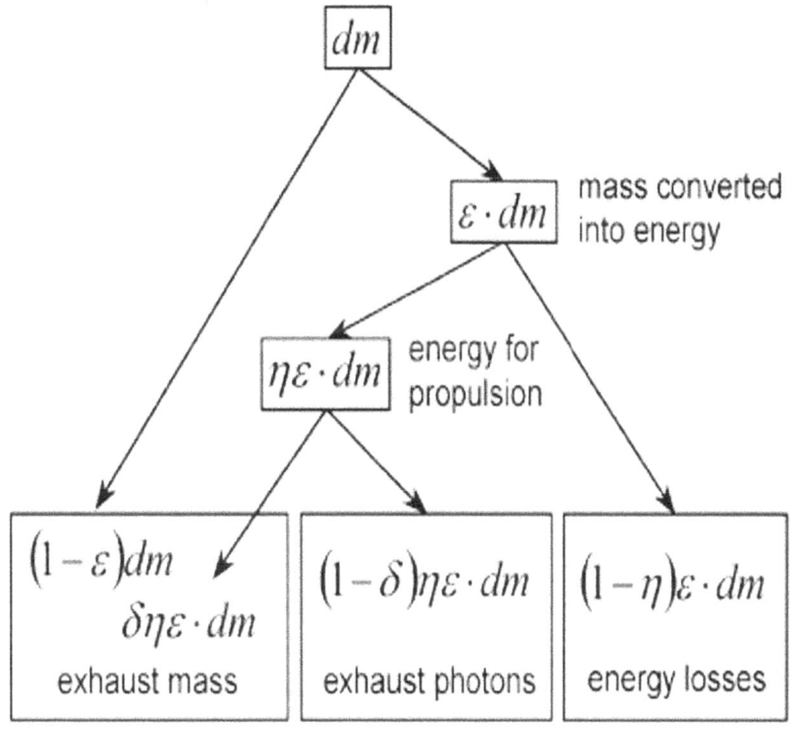

Fig 8.1:   Energy scheme for a relativistic rocket with energy losses and expelled propulsion mass and photons.

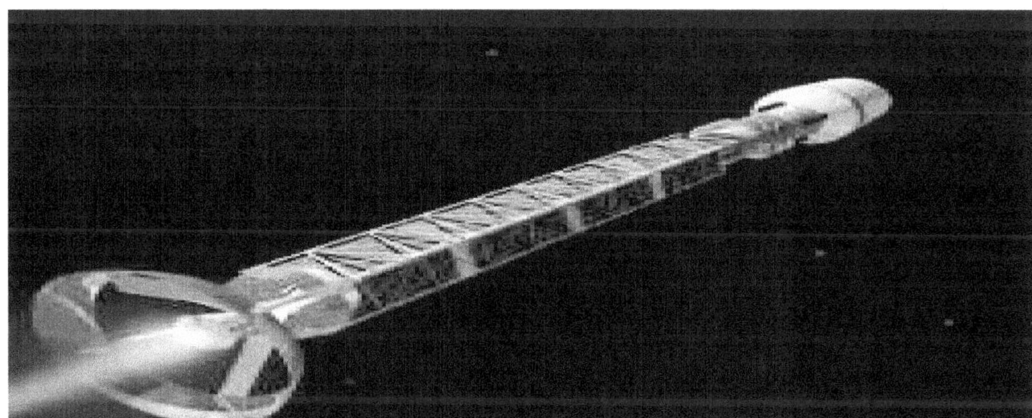

Fig 8.2: Hypothetical rocket design

9. HIGH-TEMPERATURE (100K) SUPERCONDUCTOR MAGNET MAGNETIC NOZZLE

For directing charged pions to get thrust magnetic nozzle is used. At the point of annihilation, the strength of magnetic field is given as B_0 and at a distance the strength of magnetic field is given as B_x and formulae are as follows

$$B_0 = \frac{\mu_0 I}{2R} \qquad\qquad eqn\ 9.$$

$$B_x = \frac{\mu_0 I R^2}{2(R^2 + x^2)^{3/2}} \qquad\qquad eqn\ 9.$$

where μ_0 = 1.256E-6 Tesla-m/Amp. The magnet cross-sectional area (A) is then:

$$A = {}^{I}\!/_{I_0} \qquad\qquad eqn\ 9.$$

and the magnet mass (M) is:

$$Mass = Density \cdot A \cdot 2pR$$

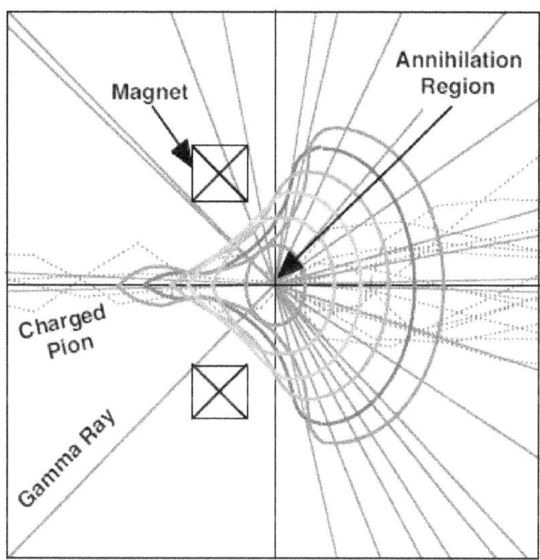

Fig 9.1: Superconducting magnetic nozzle

10. <u>PROPELLANT STORAGE AND FEED SYSTEM</u>

The normal matter hydrogen is stored as ordinary liquid hydrogen (LH2) and the antimatter as solid anti-H2 (anti-SH2). Note that the anti-SH2 must be stored and fed into the engine using a noncontact magnetic levitation technique. The anti-SH2 is stored as pellets to allow removal and transport of small quantities of antimatter to engine. For comparison, a 1-mm diameter (4.6 µg mass) anti-SH2 pellet has an m annihilation energy equivalent to 2 tons of TNT. However, storage of antiSH2 as individual pellets will result in a reduction in its effective storage density. We assumed an effective antiSH2 density 1/10 that of liquid H2 (1/12.57 of solid hydrogen)

The tank mass is calculated on the assumption of a cylindrical tank (whose diameter is fixed by the superconductor magnet diameter) with hemispherical end-domes. The required tank size is first found based on the volume of propellant required plus any tank usage (i.e., the volume of vapor above the liquid). With the tank dimensions known, the tank mass is calculated using a tank wall thickness of O.5 mm. For comparison, the Centaur LO2/LH2 chemical stage has tank walls of 1O mil; a soda pop can has a wall thickness of 3 mil. We assumed that a minimal tank pressure (e.g., 30 psia LH2) would allow the use of thin tank walls. However, it may be necessary to substantially increase the wall thickness if the propellant tanks are used as vehicle structure.

Note that the sizing of the propellant feed systems represents a completely arbitrary assumption;

This is an area that will need much more study to evaluate options and determine their mass and power requirements. For example, we have identified two options for consideration in future studies. In the first, the anti-SH2 pellets are fed magnetically down a long tube (1OOs of km in length at 1K next to a 15OOK radiator!) in to the engine. A second option would be to first convert the anti-SH2 into antiprotons, and feed the antiproton ions down to the engine in what is basically a particle beam. This option was assumed in this study.

To do this, the solid must be vaporized, ionized, and the ions accelerated to some significant speed in the particle beam. This requires electrical energy that would be supplied by the electric power system. For these calculations, we assumed a particle beam ion (antiproton) velocity of O.1%c, which gives an energy requirement of 4.65x $\boxed{10}$ J/kg. Combining this with an assumed electrical efficiency of 90% and a propellant mass flow rate gives the electric power requirement. Note however that one disadvantage of this approach is that the ion space-charge limit would result in the requirement of a very large diameter beam tube because of the large propellant flow rate needed. For example, with a space charge limit of 1O1O ions/cm3, the tube diameter would be almost three times larger than the diameter of the 4th stage of the nominal vehicle described below. However, this effect was ignored in these analyses.

11. NUCLEAR PULSE PROPULSION

Nuclear pulse fusion propulsion can provide a near term method for propelling vehicles toward the stars. The ban on nuclear weapons in space does not allow a critical mass of plutonium or uranium to be used to drive a fusion reaction. Lewis, et al, (Lewis, 1990) have shown that the fission reaction can be driven by annihilating antiproton sin plutonium or uranium nuclei. The fusion products can be used to drive a fusion reaction. The deuterium-tritium fusion has the lowest ignition temperature and proceeds according to

$$_1^2H + _1^3H \rightarrow _2^4He + _0^1$$

The helium nucleus has a kinetic energy of 3.5 MeV and the neutron has a kinetic energy of 14 MeV. The energy of the neutron is difficult to use but the helium nuclei can be readily directed by magnetic fields, or its kinetic energy can be deposited in a fluid or solid. In (Cassenti, 1997), a containment shell of tungsten surrounds a pellet consisting of deuterium and tritium. Antiprotons are sharply focused on a small chip of uranium. The sharp focus produces magnetic fields, which help to contain the fission fragments, and the fission fragments initiate the fusion reaction. The outer tungsten shell also helps to contain the fusion reaction products. An estimate of the specific impulse can be readily obtained by considering the energy released. The total energy released, E is given by

$$E_T = N_{He}E_{He} = N_DE$$

Where N is the number of helium nuclei, which is equal to the number of deuterium nuclei consumed, N_D and E is the kinetic energy of the helium nuclei (3.5 MeV). The number of deuterium nuclei, N

$$N_D = \frac{1}{2}\left[\frac{\frac{4}{3}\pi\rho_D R_D{}^3}{M_D M_{amu}}\right] \qquad eqn\ 11.$$

where lad is the density of deuterium (0.255g/cm3), RD is the radius of the deuterium-tritium fuel (1 cm). RD is one centimeter in Cassenti, 1997). M o is the atomic weight of deuterium (M D is 2), Mare u is the atomic mass unit (931.5MeV/c 2) and c is the speed of light. The factor of one-half is due to the fact that only half of the atoms are deuterium. Since the mass defect due to fusion is small, the mass before and after the fusion reaction is very nearly the same. We can now calculate the velocity, v e, of the pellet of total mass, m_t as

$$E_T = {}^1\!/_2\, m_T V_e{}^2 \qquad eqn11.$$

In (Cassenti, 1995), the pellet mass is 3.5g for a tungsten shell 0.L mm thick. The specific impulse can now be found from

$$I_{sp} = {V_e}\!/{g_0} \qquad eqn\ 1\}$$

where go is the acceleration of gravity at the surface of the Earth.

12. <u>PERFORMANCE SIZING</u>

The mission design criteria are the distance to be covered, s_t, and the time to complete the mission, t, All missions to be considered will be fly-by missions. If the time the engine fires, t_f, is much less than the mission time, then the final speed required is simply

$$V_f = {S_t}/{t_t} \qquad \textit{eqn 12.}$$

The mass ratio, MR, given by the ratio of initial mass, mi, and the final mass, m is the

$$MR = {m_i}/{m_f} = e^{V_f/V_e} \qquad \textit{eqn 12.}$$

Where $\quad$ is the exhaust velocity and is given by

$$V_e = I_{sp}g_0 \qquad \textit{eqn 1}$$

and $\dot{\eta}$ is the fraction of energy actually released. The product of the mass flux, $\quad$ and the exhaust velocity gives the thrust, F. This flux is known from the pellet injection frequency, N, and the pellet mass. Then

$$F = \dot{N}m_T V_e \qquad \textit{eqn 12.}$$

$$F = \dot{N}\sqrt{2\eta E_T m_T} \qquad \textit{eqn 12.}$$

The mass flux is also

$$\dot{m} = \frac{m_i - m_f}{t_f} = \frac{F}{V_e} \qquad \textit{eqn 12.}$$

Then

$$t_f = \frac{(m_i - m_f)V_e}{F} \qquad \textit{eqn 12.}$$

is the time the rocket fires.

If ☐ is much less than ☐ then the estimates given above are accurate. If not, the distance traveled while the rocket fires must be found. The mass at any time during firing is given by

$$m = m_i e^{-V/V_e} = m_i - \frac{F_t}{V_e} \qquad \textit{eqn 12.}$$

Solving for the speed

$$V = V_e \ln \left[\frac{1}{1 - \frac{Ft}{m_i V_e}} \right] = \frac{ds}{dt} \qquad \textit{eqn 12.}$$

where s is the distance traveled, and t is time.

Integrating and noting s is zero at t equal to zero, the distance traveled in time ☐ is

$$S_f = \frac{m_i V_e^2}{F} \left[1 - \frac{m_f}{m_i} + \frac{m_f}{m_i} \ln \frac{m_f}{m_i} \right] \qquad \textit{eqn 12.1}$$

An iterative solution can now be performed using a new value for ☐ given by

$$V_f = \frac{S_T - S_f}{t_T - t_f} \qquad \textit{eqn 12.1}$$

13. <u>SOLAR GRAVITY LOSS</u>

The speed gained during the propulsive maneuver will be smaller than predicted, due to the effects of solar gravity. An upper limit on the loss can be obtained by considering the rocket to thrust radically away from the Sun. Since the solar accelerations are about O g's and the thrust-to-mass ratio of the engine is more than O.O5Og's, the rocket will climb away from the Sun. For this case, the equation of motion is

$$m\frac{d^2r}{dt^2} = F - \frac{GMm}{r^2} \qquad eqn\ 13.$$

Where r is the distance from the Sun, M is the solar mass and G is the gravitational constant. Assuming the free acceleration, of the rocket, $\boxed{}$ is constant. The thrust is:-

$$F = ma_0 \qquad eqn\ 13.$$

Substituting in above equation we get

$$\frac{d^2r}{dt^2} = a_0 - \frac{GM}{r^2} \qquad eqn\ 13.$$

For the initial conditions, take

$$r = r_0, \frac{dr}{dt} = 0\ at\ t =$$

Integrating once

$$v = \frac{dr}{dt} = \sqrt{2a(r - r_0) - 2GM\left[\frac{1}{r_0} - \frac{1}{r}\right]} \qquad eqn\ 13.$$

If there were no solar gravity (i.e., GM = 0)

$$V_0 = \sqrt{2a(r - r_0)} \qquad \text{eqn 13.}$$

The speed loss relative to the speed with no solar gravity is

$$\frac{\Delta V}{V_0} = \frac{V_0 - V}{V_0} = \sqrt{1 - \frac{GM}{a_0 r_0{}^2}\left(\frac{r_0}{r}\right)} - 1 \qquad \text{eqn 13.}$$

For small values of

$$\frac{GM}{a_0 r_0{}^2}\left(\frac{r_0}{r}\right) \qquad \text{eqn 13.}$$

$$\frac{\Delta V}{V} = \frac{GM}{2 a_0 r_0{}^2}\left(\frac{r^0}{r}\right) \qquad \text{eqn 13.}$$

If the final desired speed is ☐ at r equal to ☐ then

$$V_f{}^2 = 2 a_0 r_f \qquad \text{eqn 13.}$$

and equation becomes at r equal to ☐

$$\frac{\Delta V}{V_f} = \frac{GM/r_0}{V_f^2/2} \qquad \textbf{eqn 13.1}$$

which is the ratio the initial solar gravitational energy to the Final kinetic energy.

Setting $\square$

$$V_f = \frac{S_T}{t_T} \qquad \textbf{eqn 13.1}$$

and noting $v\,o \sim \square$:

$$\frac{\Delta V}{V_f} = \frac{2GM t_T^2}{S_T^2 r_0} \qquad \textbf{eqn 13.1}$$

14. <u>RELATIVISTIC EFFECTS</u>

When we talk about slow speeds then the classical mechanics or the Newtonian mechanics gives us the desired result but when we are discussing speeds close to speed of light or any fraction of it then the relativistic effects must be considered. For slow speeds Newtonian mechanics coincides with the result given by relativity, but because of its simpler nature Newtonian mechanics is used.

The relativistic effects on a body on its mass, length and time are given by the Lorentz factor

$$\beta = v \quad \text{and} \quad \gamma = \frac{1}{\sqrt{1-}}$$

The relativistic effects acting on a body moving with a speed close to light are mass dilation, length contraction and time dilation.

14.1. Mass Dilation

Mass dilation is nothing but the variation or the significant increase in mass when a body moves with the velocity close to the speed of light. The mass of the object in reference frame is called rest mass. The reference frame is nothing but the frame with respect to observer or in simpler words the frame in with the observer is standing.

For example, if a body weighs 10kg at rest, or more properly when measured in a reference, then this mass appears to be less than the mass when the same object moves with a velocity close enough to the speed of light. To get a measurable difference in mass the speeds must be very high.

The equation for calculating the mass as seen by an outside observer is as follows:

$$m = \frac{m_0}{\sqrt{1 - \frac{v^2}{c^2}}}$$

Where: m_0 = the mass measured at rest relative to an observer traveling with the same velocity as the mass, the "rest mass".

m = the mass measured by the observers on the other reference frame.

v = the speed of the object

c = the speed of light in a vacuum

<u>**Mass increase due to very high speeds**</u>

Mass at 0% of the
speed of light = 10kg

Mass at 99.5 % of the
speed of light = 100kg

14.2. Length Contraction

The phenomenon of decrease in length as observed when objects are travelling at any non -zero velocity relative to observer. This contraction is usually noticeable at a significant fraction of speed of light. Contraction is only in the direction parallel to the direction in which observed body is travelling. The effect is negligible for regular speeds and becomes relevant at greater speeds. As the magnitude of the velocity approaches the speed of light, the effect becomes dominant, as given by the formula

$$L = \frac{L_0}{\gamma(v)} = L_0 \sqrt{1 - \frac{v^2}{c^2}}$$

where

    L_0 is the proper length (the length of the object in its rest frame),

    L is the length observed by an observer in relative motion with respect to the object,

    v is the relative velocity between the observer and the moving object,

    c is the speed of light,

and the *Lorentz factor*, $\gamma(v)$, is defined as

$$\gamma(v) = \frac{1}{\sqrt{1 - \frac{v^2}{c^2}}}$$

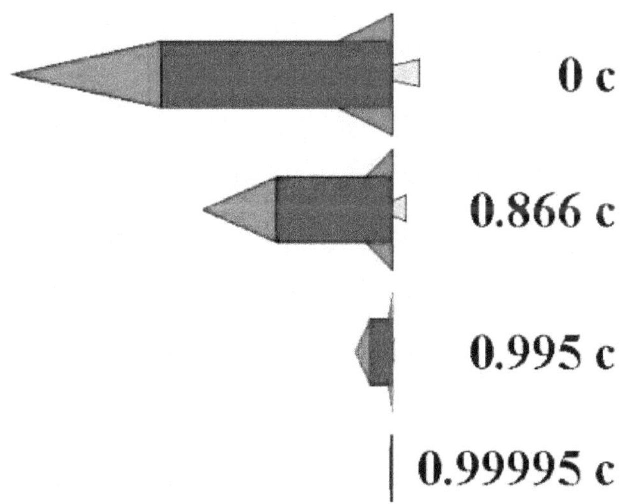

Fig 14.2 relativistic length contraction

14.3. Time Dilation

The phenomenon of time dilation is the actual difference between elapsed times observed by the observers moving relative to each other.

A clock at rest with respect to one observer may be measured to tick at a different rate when compared to a second observer's own equally accurate clocks. The effect arises neither from technical aspects of the clocks nor from the fact that signals need time to propagate, but from the nature of space time itself. The time dilation due to relative velocity can be determined by the formula-

$$\Delta t' = \gamma \Delta t = \frac{\Delta}{\sqrt{1-}}$$

where Δt is the time interval between *two co-local events* (i.e. happening at the same place) for an observer in some inertial frame (e.g. ticks on his clock), this is known as the *proper time*, $\Delta t'$ is the time interval between those same events, as measured by another observer, initially moving with velocity v with respect to the former observer, v is the relative velocity between the observer and the moving clock, c is thespeed of light, and the Lorentz factor

$$\gamma = \frac{1}{\sqrt{1-\frac{1}{c}}}$$

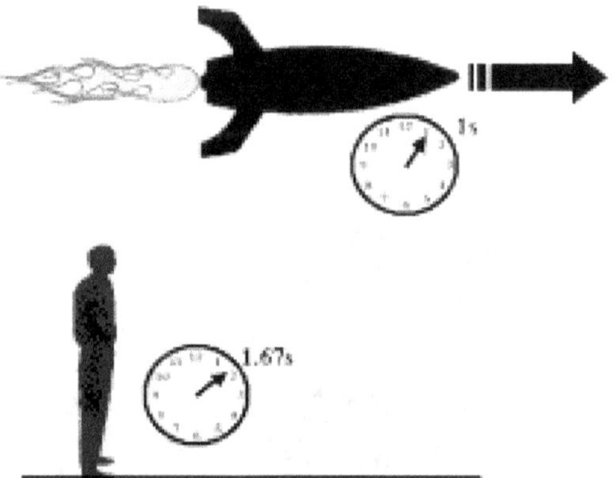

Fig 14.3: Relativistic time dilation

15. <u>MISSION ANALYSIS</u>

15.1. Mission Analysis for Proxima Centauri

15.1.1. Variation in Actual time and dilated time with respect to the distance

Distance (LY)	Δ (yrs)	$\Delta t (yr$
1.00	10.0	10.05
2.00	20.0	20.10
3.00	30.0	30.15
4.00	40.0	40.20
4.26	42.6	42.81

Δt = dilated time measured from the Earth

Δ = proper time or the time measured by the clocks on the spacecraft itself

15.1.2. Variation in velocity due to solar gravity

Distance(AU)	Velocity(m/s)
10000	1.20E+07
20000	1.69E+07
30000	2.07E+07
40000	2.39E+07
50000	2.67E+07

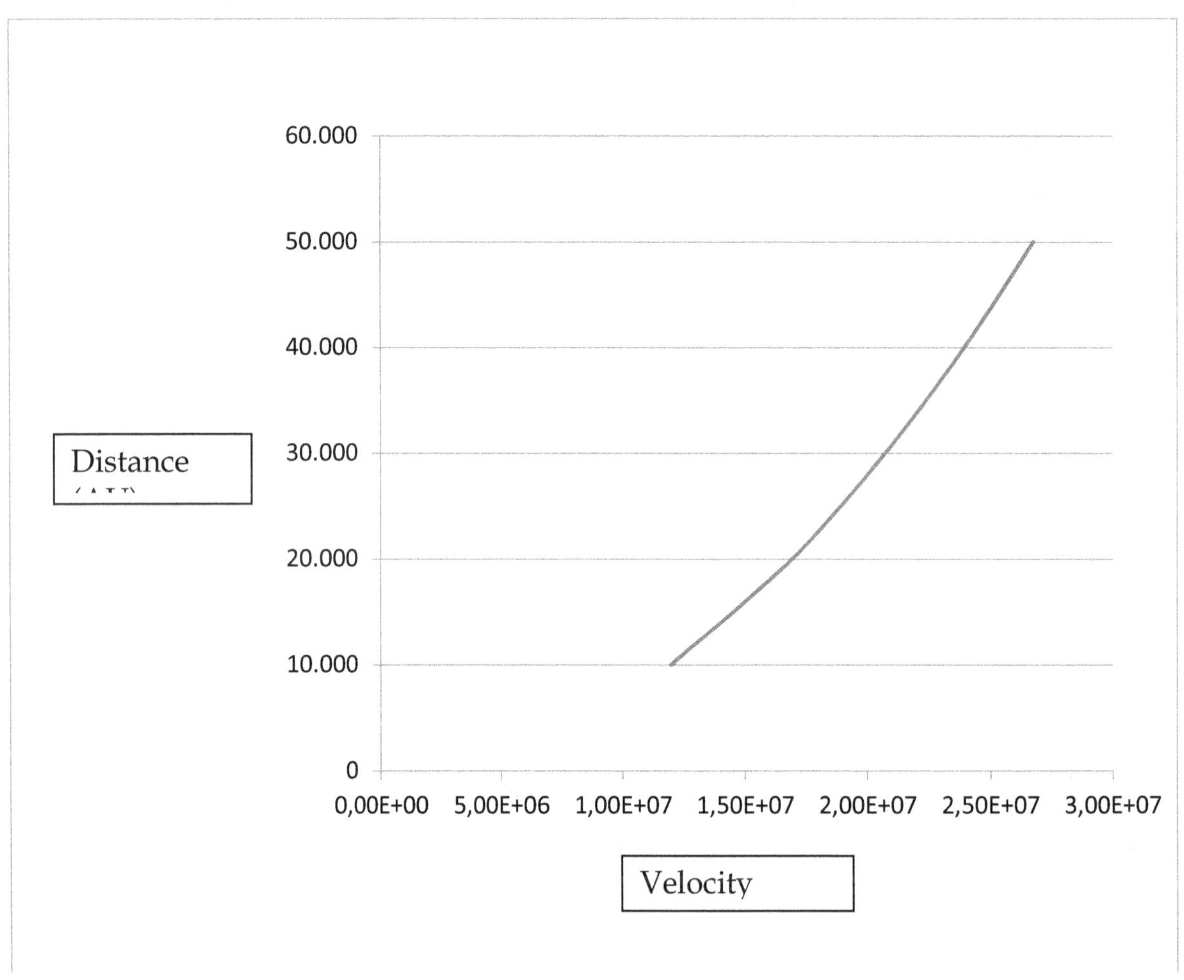

Fig 15.1: Graph showing Variation in velocity due to solar gravity

15.1.3. Velocity loss relation in absence of solar gravity

Distance (LY)	$\Delta V / V$
1.OO	2.01E-06
2.OO	2.01E-06
3.OO	2.01E-06
4.OO	2.01E-06
4.26	2.01E-06

Now the graphical representation

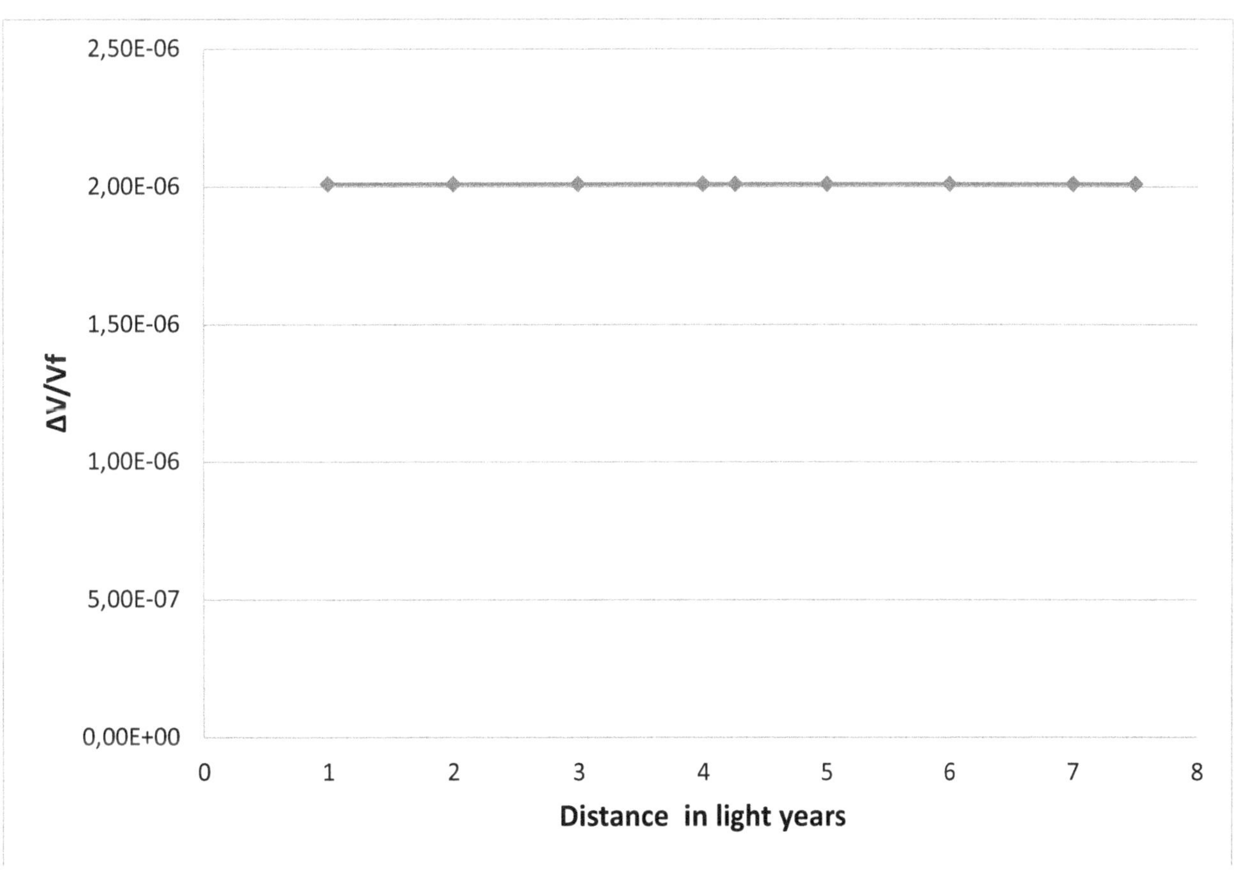

Fig 15.2: Graph showing Velocity loss relation in absence of solar gravity

15.1.4. Relativistic effect of mass

Due to the relativistic effect on mass we have plot a table to demonstrate the mass variation at different distances covered by the spacecraft

Distance (AU)	M/M
10000	1.0008
20000	1.0016
30000	1.0024
40000	1.0032
50000	1.0040

And the graph representation

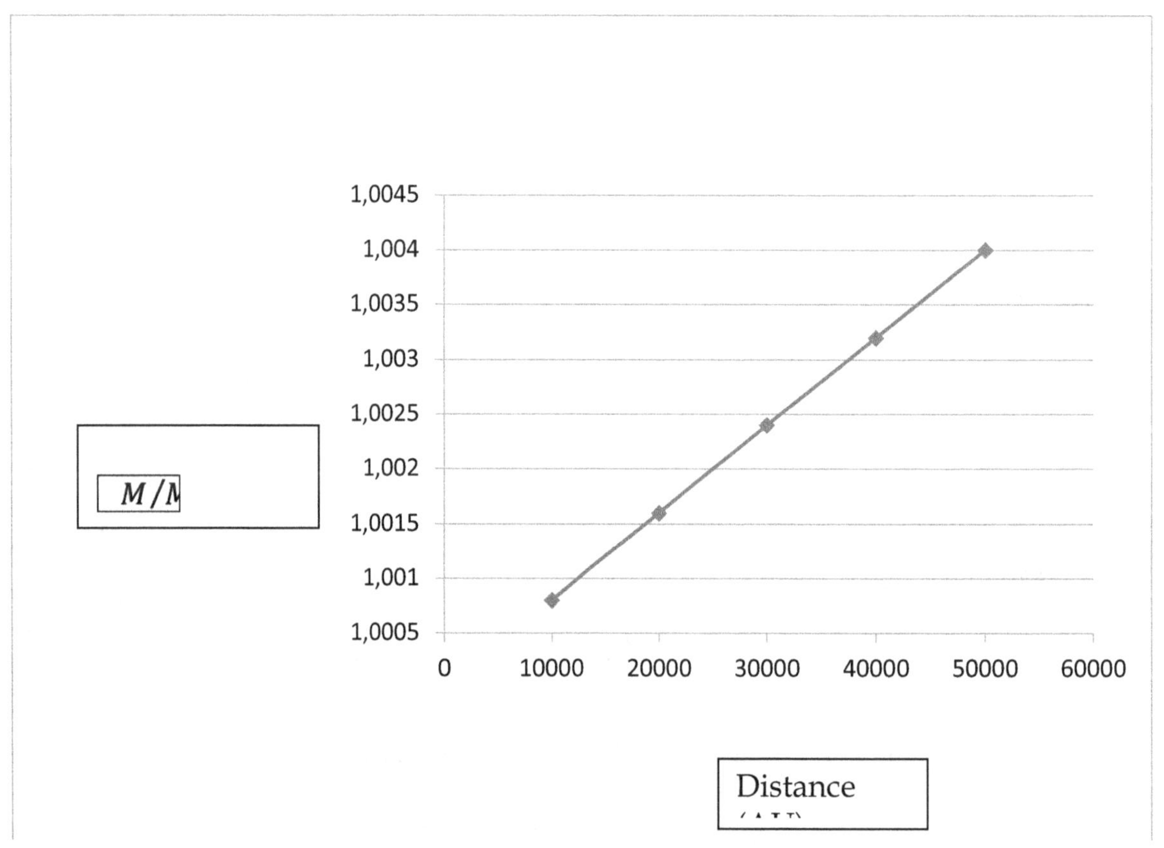

15.2. Mission Analysis for Barnard Star

15.2.1. Variation in Actual time and dilated time with respect to the distance

Distance (LY)	$\Delta\tau$ (yrs)	Δt (yrs)
1.00	10.0	10.05
2.00	20.0	20.10
3.00	30.0	30.15
4.00	40.0	40.20
4.26	42.6	42.81
5.00	50.0	50.25
6.00	60.0	60.30

Δt = dilated time measured from the Earth

$\Delta\tau$ = proper time or the time measured by the clocks on the spacecraft itself

15.2.2. Variation in velocity due to solar gravity

Distance(AU)	Velocity(m/s)
10000	1.20E+07
20000	1.69E+07
30000	2.07E+07
40000	2.39E+07
50000	2.67E+07

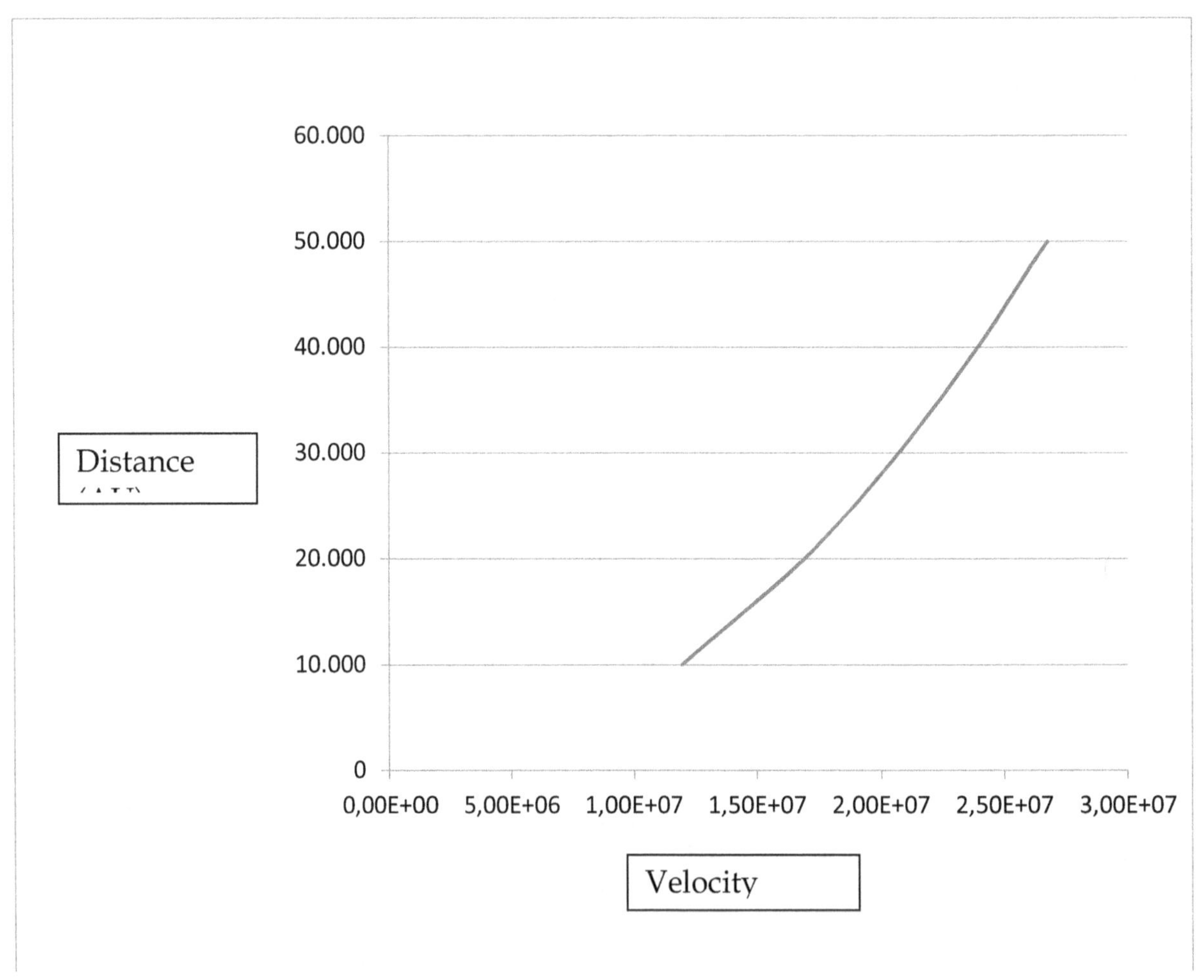

Fig 15.4: Graph showing variation in velocity due to solar gravity

15.2.3. Velocity loss relation in absence of solar gravity

Distance (LY)	$\Delta V / V_f$
1.00	2.01E-06
2.00	2.01E-06
3.00	2.01E-06
4.00	2.01E-06
4.26	2.01E-06
5.00	2.01E-06
6.00	2.01E-06

Now the graphical representation

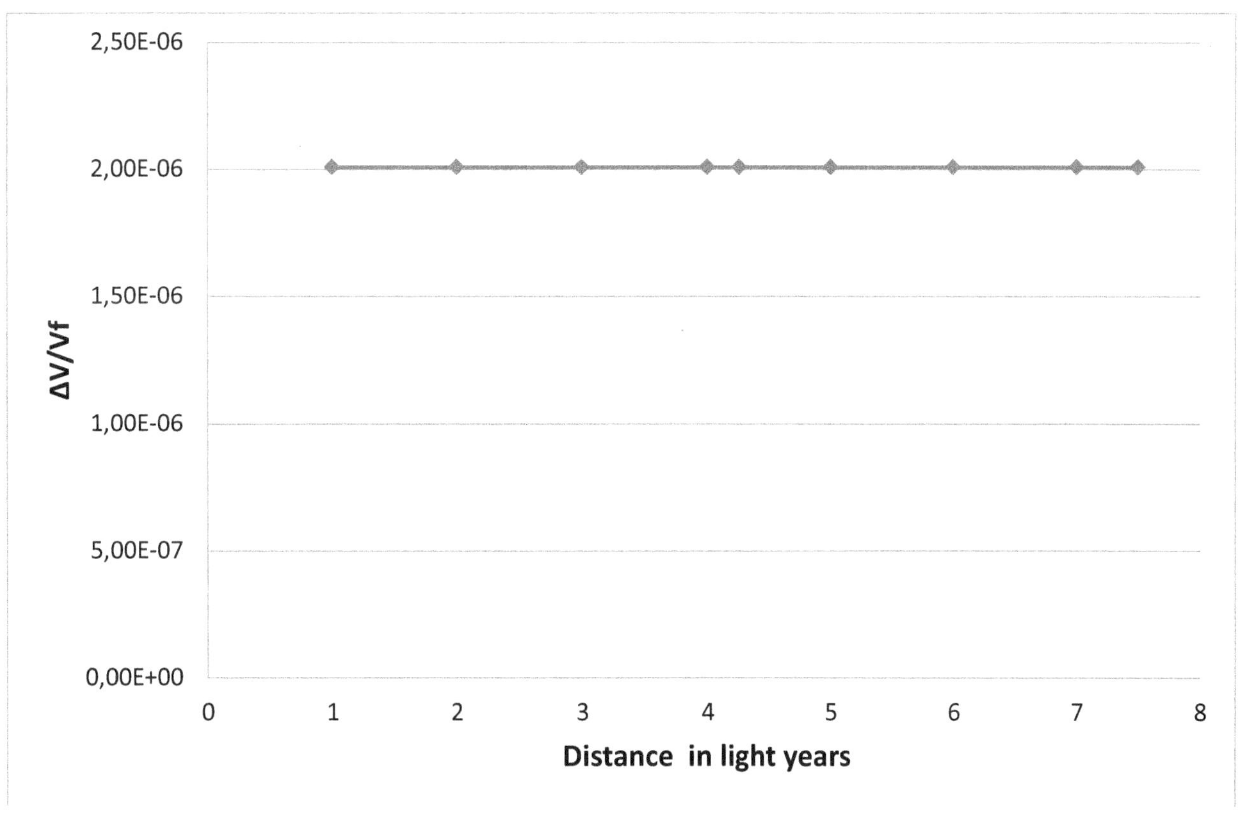

Fig 15.5: Graph showing Velocity loss relation in absence of solar gravit

15.2.4. Relativistic effect of mass

Due to the relativistic effect on mass, we have plot a table to demonstrate the mass variation at different distances covered by the spacecraft

Distance (AU)	M/M
10000	1.0008
20000	1.0016
30000	1.0024
40000	1.0032
50000	1.0040

And the graph representation

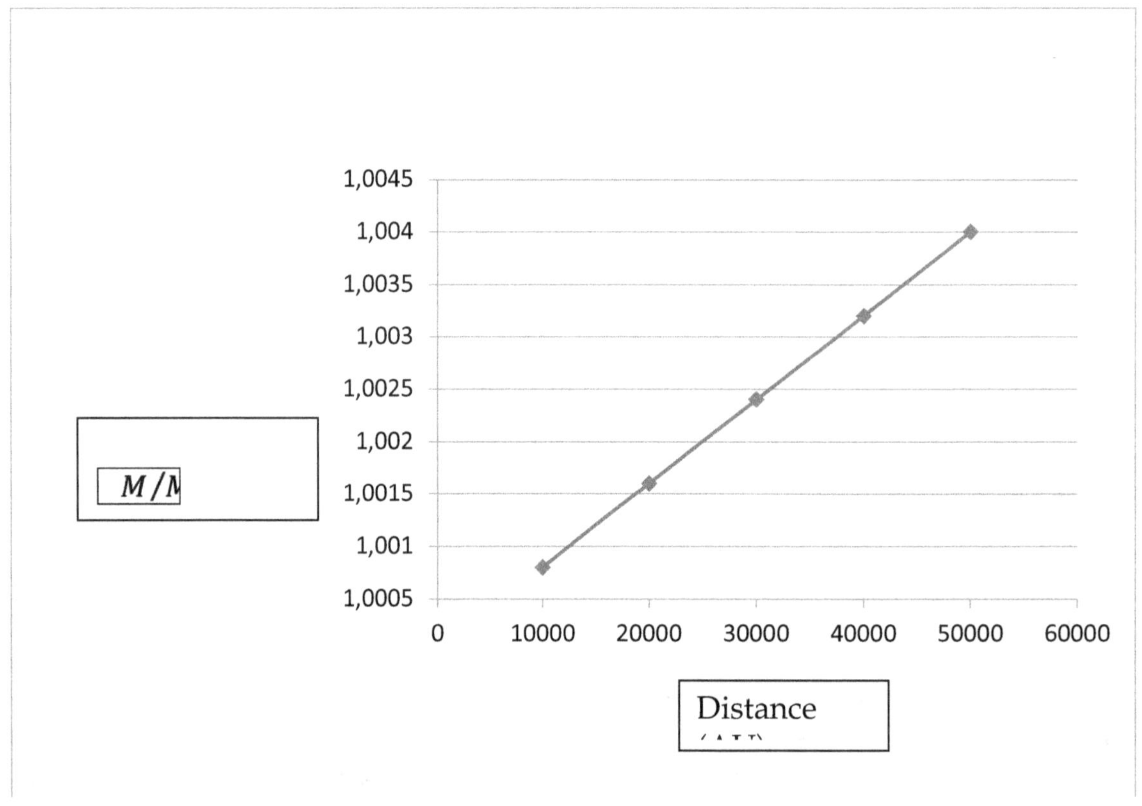

Fig 15.6: Graph showing relativistic variation in mass

15.3.1. Variation in Actual time and dilated time with respect to the distance

Distance (LY)	$\Delta\tau$(yrs)	$\Delta t(yrs)$
1.00	10.0	10.05
2.00	20.0	20.10
3.00	30.0	30.15
4.00	40.0	40.20
4.26	42.6	42.81
5.00	50.0	50.25
6.00	60.0	60.30
7.00	70.0	70.35
7.50	75.0	75.38
7.70	77.0	77.39

Δt= dilated time measured from the Earth

$\Delta\tau$ = proper time or the time measured by the clocks on the spacecraft itself

15.3.2. Variation in velocity due to solar gravity

Distance(AU)	Velocity(m/s)
10000	1.20E+07
20000	1.69E+07
30000	2.07E+07
40000	2.39E+07
50000	2.67E+07

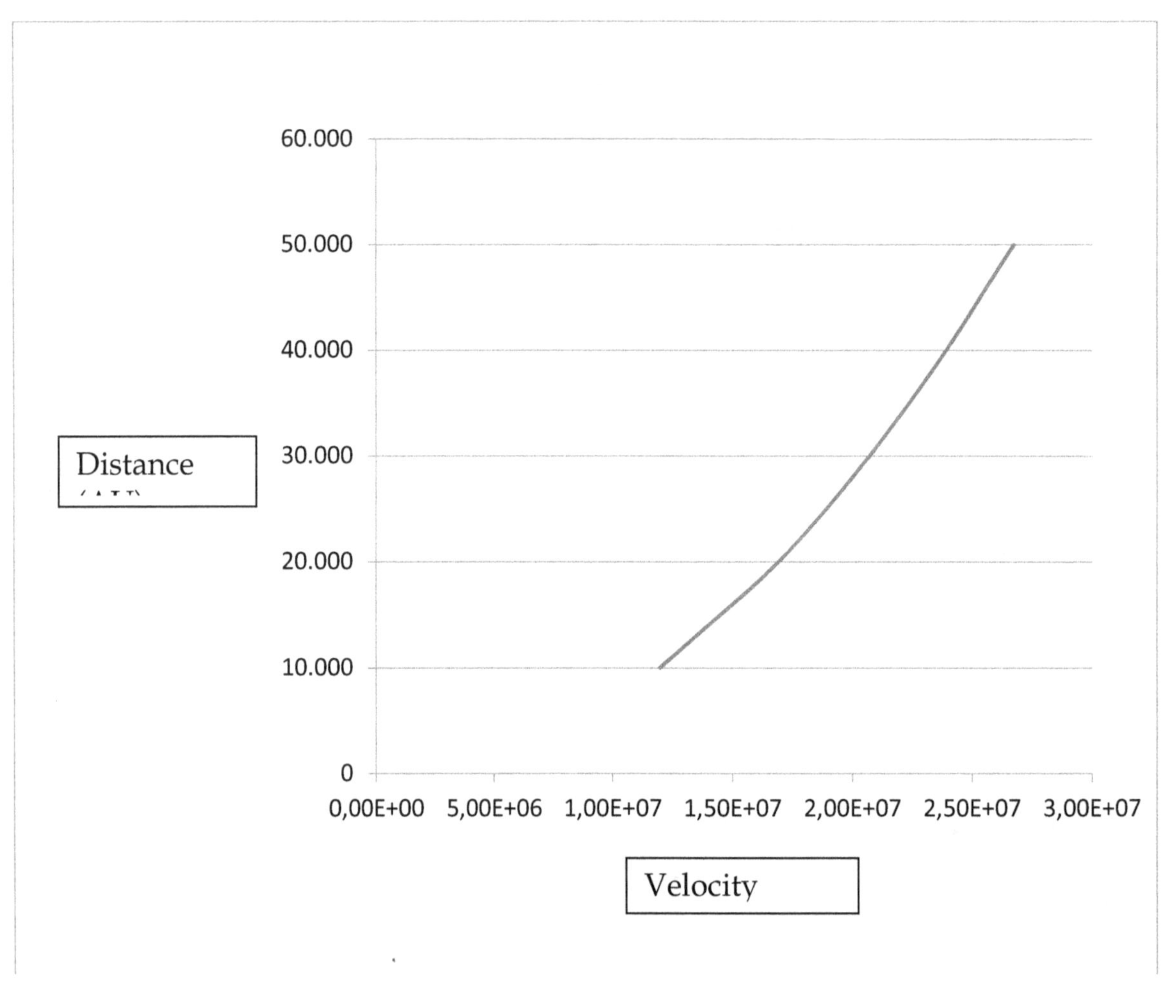

Fig 15.7: Graph showing variation in velocity due to solar gravity

15.3.3. Velocity loss relation in absence of solar gravity

Distance (LY)	$\Delta V / V_f$
1.00	2.01E-06
2.00	2.01E-06
3.00	2.01E-06
4.00	2.01E-06
4.26	2.01E-06
5.00	2.01E-06
6.00	2.01E-06
7.00	2.01E-06
7.50	2.01E-06
7.70	2.01E-06

Now the graphical representation

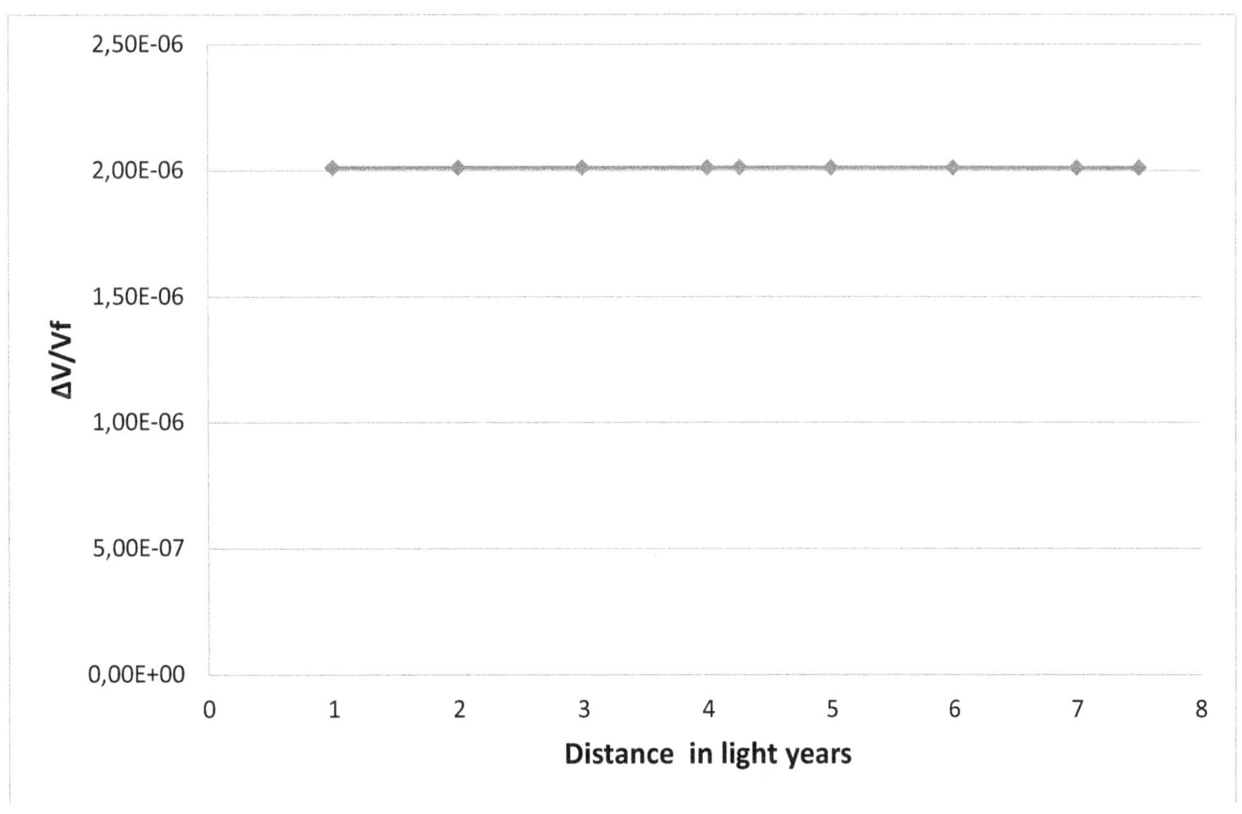

Fig 15.8: Graph showing Velocity loss relation in absence of solar gravit

15.3.4. Relativistic effect of mass

Due to the relativistic effect on mass we have plot a table to demonstrate the mass variation at different distances covered by the spacecraft

Distance (AU)	M/M
10000	1.0008
20000	1.0016
30000	1.0024
40000	1.0032
50000	1.0040

And the graph representation

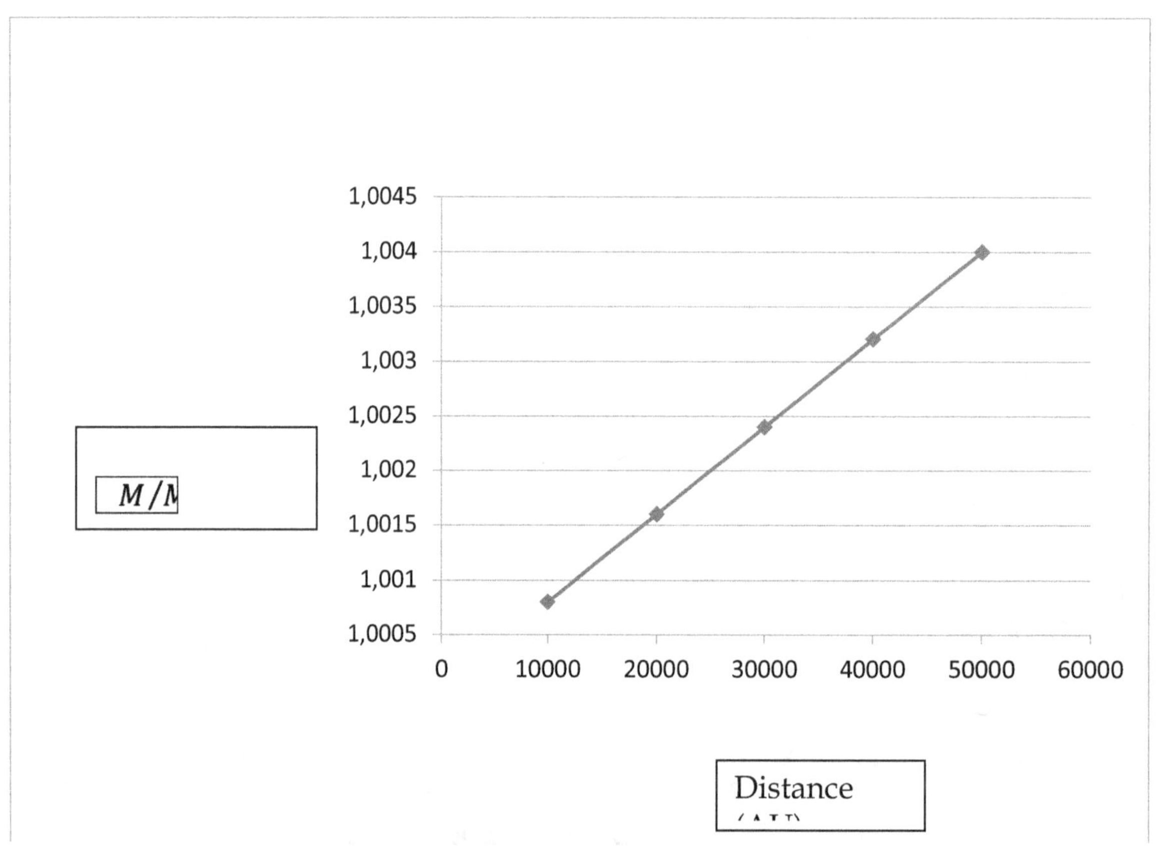

Fig 15.9: Graph showing relativistic variation in mass

15.4.1. Variation in Actual time and dilated time with respect to the distance

Distance (LY)	Δ(yrs)	$\Delta t(yr)$
1.00	10.0	10.05
2.00	20.0	20.10
3.00	30.0	30.15
4.00	40.0	40.20
4.26	42.6	42.81
5.00	50.0	50.25
6.00	60.0	60.30
7.00	70.0	70.35
7.50	75.0	75.38
7.70	77.0	77.39
8.00	80.0	80.40
8.20	82.0	82.41

Δt= dilated time measured from the Earth

Δ = proper time or the time measured by the clocks on the spacecraft itself

15.4.2. Variation in velocity due to solar gravity

Distance(AU)	Velocity(m/s)
10000	1.20E+07
20000	1.69E+07
30000	2.07E+07
40000	2.39E+07
50000	2.67E+07

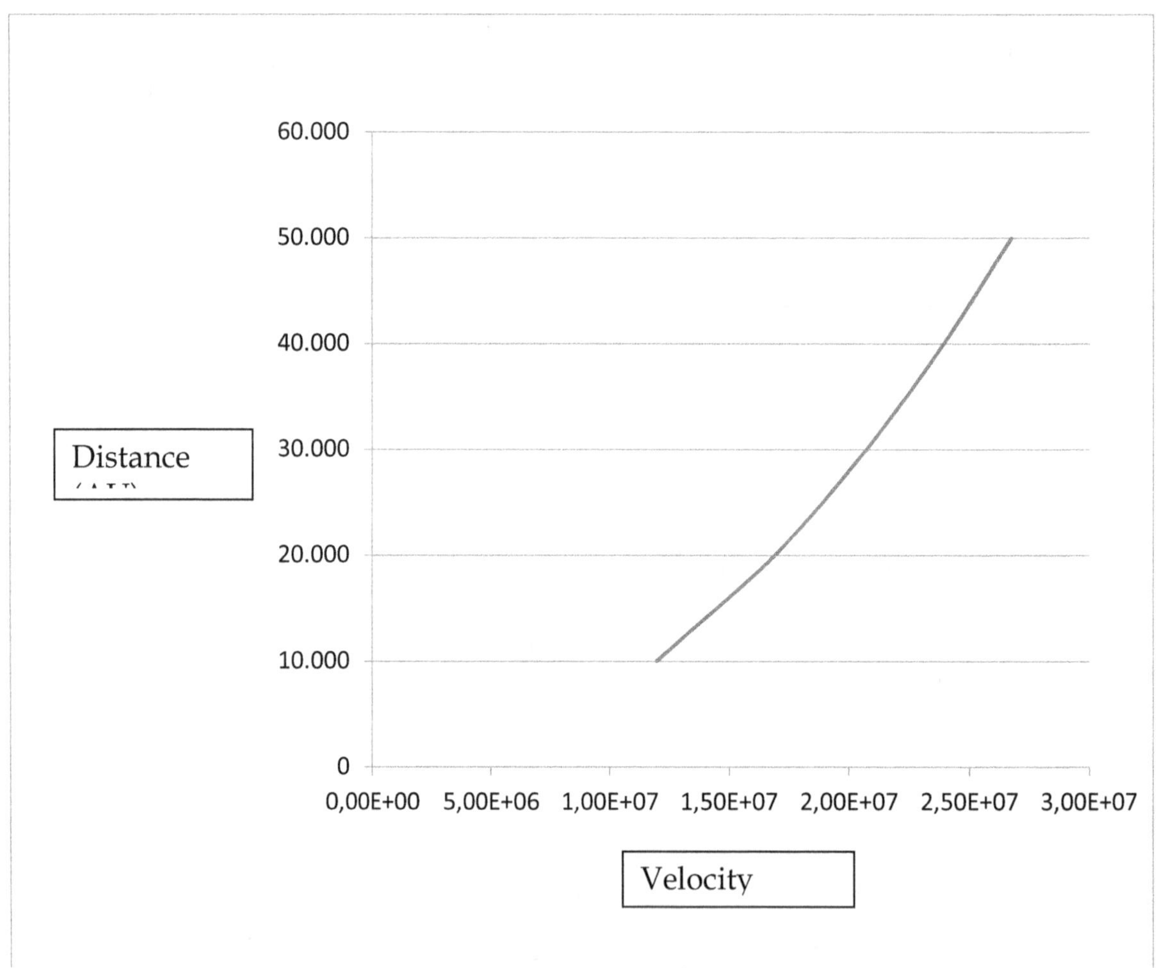

Fig 15.10: Graph showing variation in velocity due to solar gravity

15.4.3. Velocity loss relation in absence of solar gravity

Distance (LY)	$\Delta V/V_f$
1.00	2.01E-06
2.00	2.01E-06
3.00	2.01E-06
4.00	2.01E-06
4.26	2.01E-06
5.00	2.01E-06
6.00	2.01E-06
7.00	2.01E-06
7.50	2.01E-06
7.70	2.01E-06
8.00	2.01E-06
8.20	2.01E-06

Now the graphical representation

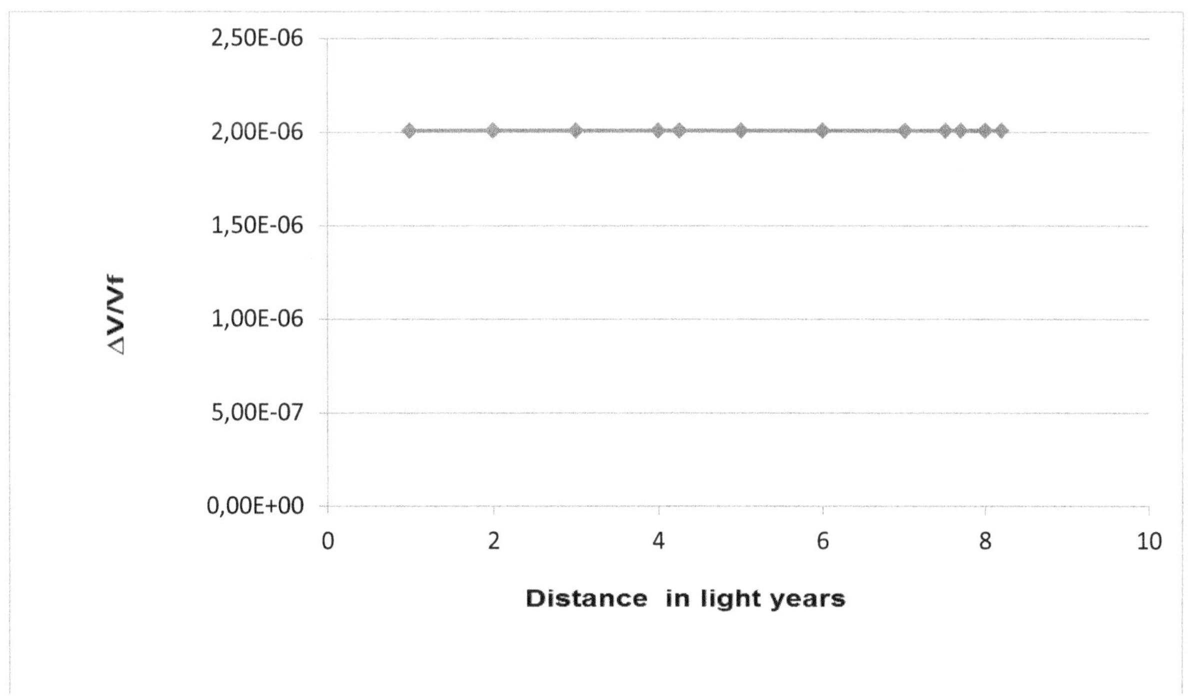

Fig 15.11: Graph showing Velocity loss relation in absence of solar gravity

15.4.4. Relativistic effect of mass

Due to the relativistic effect on mass, we have plot a table to demonstrate the mass variation at different distances covered by the spacecraft

Distance (AU)	M/M
10000	1.0008
20000	1.0016
30000	1.0024
40000	1.0032
50000	1.0040

And the graph representation

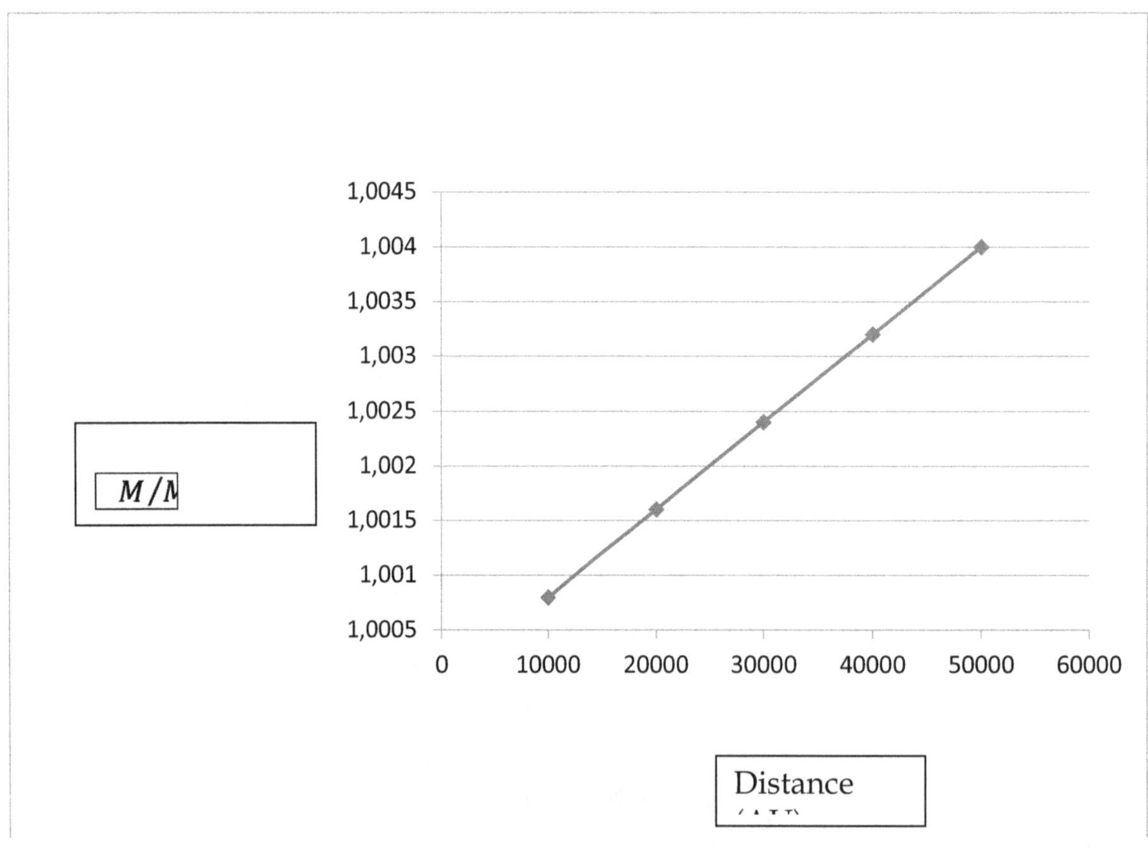

Fig 15.12: Graph showing relativistic variation in mass

16. PROGRAMMING

16.1. For Solar Gravity loss

```
program solarloss
real ::a=0.048                  !acceleration
real ::GM=1.334E+20             !G-gravitational constant and M-mass of Sun
real ::r=1                 !distance between Sun and earth (1 AU)
real v                     !velocity at the required location ( m/s)
real Z                     !distance to find the velocity at a particular location (in
AU)
print*,"what is the value of Z"
read*,Z
v=((2*a*(Z-r)*149E+9)-(2*GM*(Z-r)/(Z*r*149E+9)))**0.5
print*,"the velocity at that distance is"
print*,v
end program solarloss
```

```
program solarloss
real ::a=0.048
real ::GM=1.334E+20
real ::r=1
real v
real Z
print*,"what is the value of Z"
read*,Z
v=((2*a*(Z-r)*149E+9)-(2*GM*(Z-r)/(Z*r*149E+9)))**0.5
print*,"the velocity at that distance is"
print*,v
end program solarloss
```

Fig 16.1: Input program for solar gravity losses

```
 what is the value of Z
30000
 the velocity at that distance is
    2.071482E+07

Press RETURN to close window...
```

Fig 16.2: Output program for solar gravity losses

16.2. For no solar gravity loss

```fortran
program nosolargravity
real ::a=1.79E+9              !value for 2GM/r
real t                       !value of time in years
real s                       !distance in light years
real b                       !velocity ratio (change in velocity to final velocity)
print*,"what is the value of t"
read*,t
print*,"what is the distance"
read*,s
b=a*(t/(s*3E+08))**2
print*,"the ratio of change in velocity to final velocity"
print*,b
end nosolargravity
```

```fortran
program nosolargravity
real ::a=1.79E+9
real t
real s
real b
print*,"what is the value of t"
read*,t
print*,"what is the distance"
read*,s
b=a*(t/(s*3E+08))**2
print*,"the ratio of change in velocity to final velocity"
print*,b
end nosolargravity
```

Fig 16.3: Input program for no solar gravity losses

```
what is the value of t
60.30
what is the distance
6
the ratio of change in velocity to final velocity
    2.008828E-06

Press RETURN to close window...
```

Fig 16.4: Output program for no solar gravity losses

<u>REFERENCES</u>

1. Frisbee, Robert H, "ANTIMATTER ROCKET FOR INTERSTELLAR MISSIONS", 39th AIAA/ASME/SAE/ASEE Joint Propulsion Conference and Exhibit20-23 July 2003, Huntsville, Alabama
2. Close, Frank "ANTIMATTER", published in 2009

3. Keane, Ronan L.,"BEAMED CORE ANTIMATTER PROPULSION", *Western Reserve Academy, 115 College Street, Hudson, Ohio 44236, USA*

4. Cassenti, B. *"A* Comparison of Interstellar Propulsion Systems," JBIS, 35, pp. 116-124, (1982).

APPENDIX A: NASA DESIGN OF ANTIMATTER SPACESHIP TO MARS

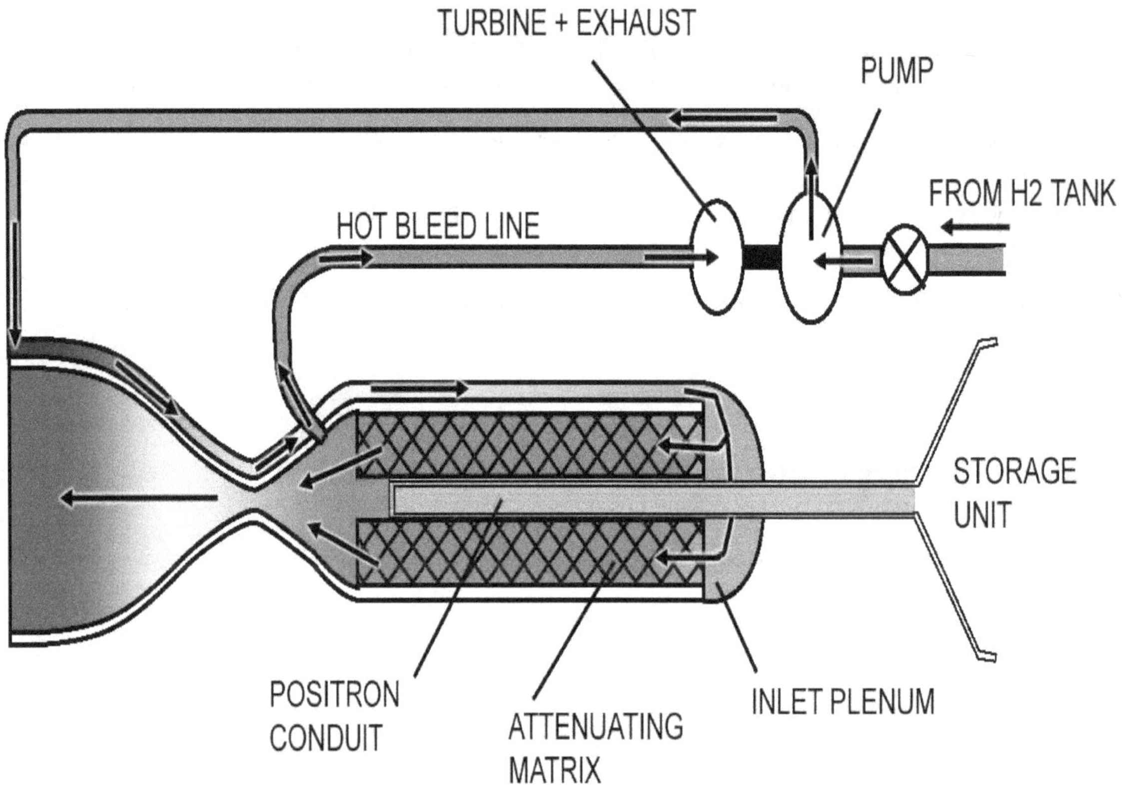

Proton-Antiproton Solid Core Engine

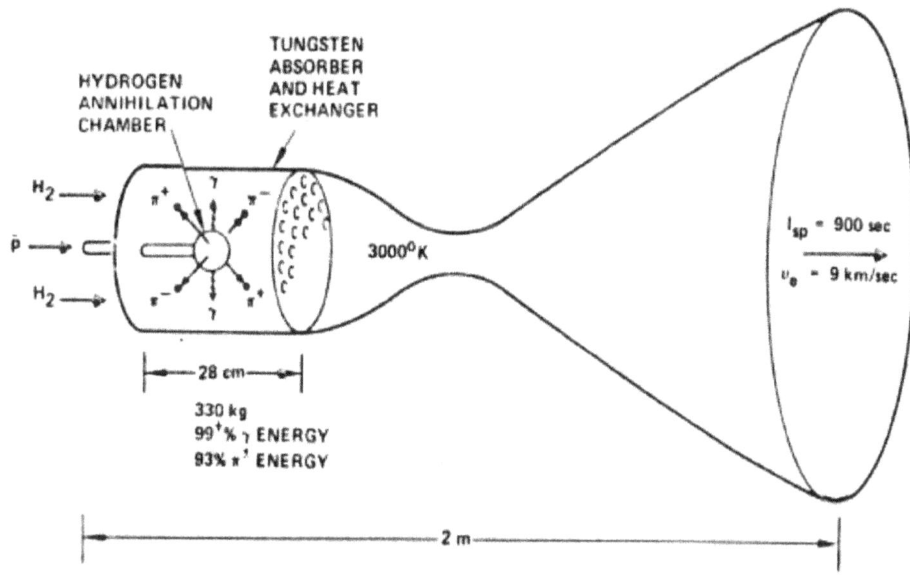

- > 90% transfer of annihilation energy to tungsten block
- Similar performance to an NTP engine (Isp ~ 900 s, high thrust)
- Typical $\bar{p}$ mass flows ~ several μg/sec (material temperature limits)

Figure: Forward, R. L., *Antiproton Annihilation Propulsion*, AFRPL TR-86-034, AFRPL/LKC, Edwards AFB, Ca., Sep. 1985.

Proton-Antiproton Gas Core Engine

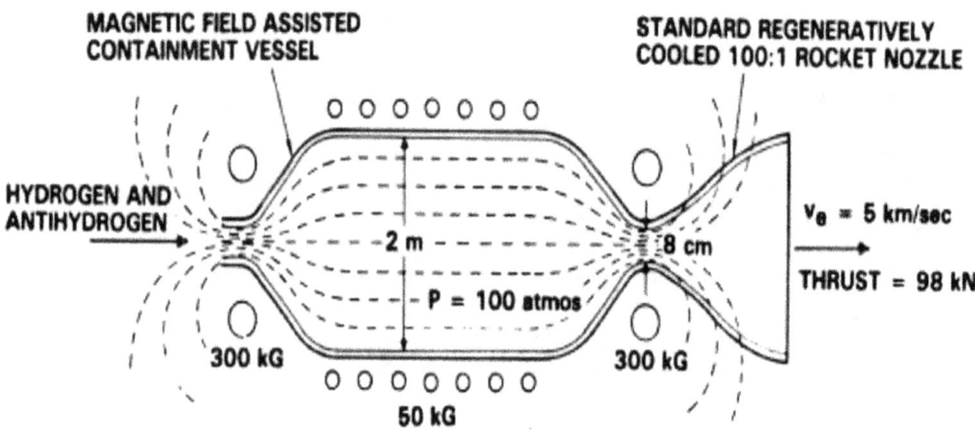

- About 35% energy transfer to the high pressure hydrogen propellant
- Specific impulse similar to chemical engines (~ 500 s), high thrust
- Variants include liquid hydrogen for better transfer efficiency
- Typical antiproton mass flow rates ~ 10's µg/sec

Figure: Forward, R. L., *Antiproton Annihilation Propulsion*, AFRPL TR-86-034, AFRPL/LKC, Edwards AFB, Ca., Sep. 1985.

Proton-Antiproton Plasma Core Engine

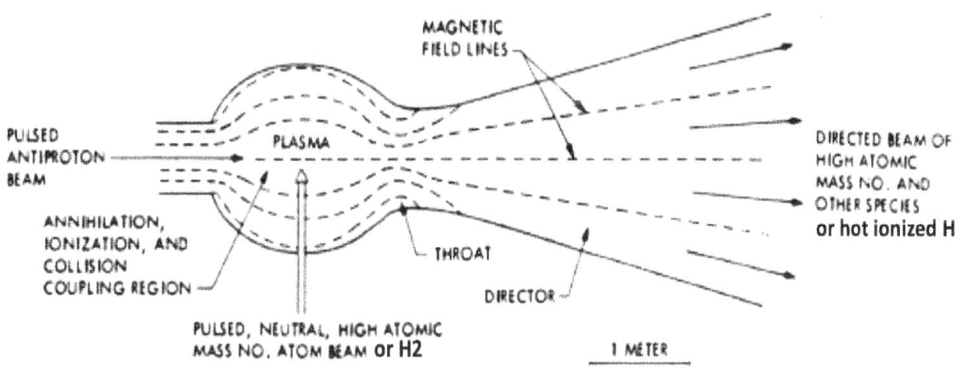

- Charged particles trapped and guided by strong magnetic fields
- Higher Isp than chemical engines (several 1000 s), moderate thrust
- Annihilation energy transferred to hydrogen is only about 1-2%*
- Typical pulse ~ 10^{18} $\bar{p}$ (depending on rep rate, ~ 100's μg/sec)
- Detailed numerical studies not yet performed for heavier elements

Figure: Forward, R. L., *Antiproton Annihilation Propulsion,* AFRPL TR-86-034, AFRPL/LKC, Edwards AFB, Ca., Sep. 1985
* LaPointe, M., "Antiproton Powered Propulsion with Magnetically Confined Plasma Engines," *J Prop & Power*, **7** (5), 1991

Proton-Antiproton Beam Core Engine

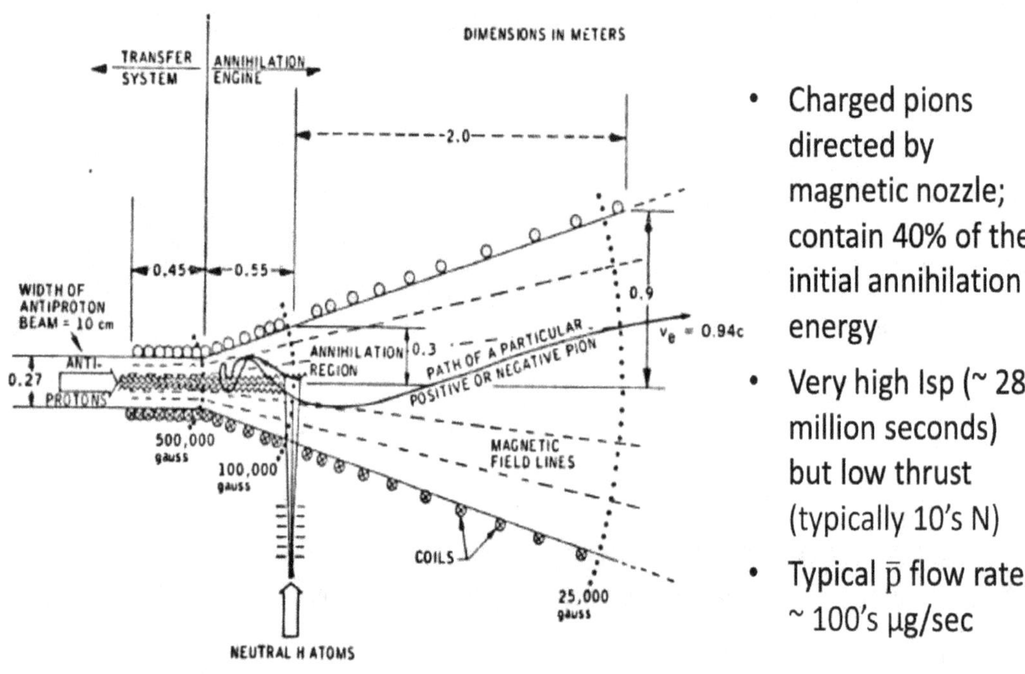

- Charged pions directed by magnetic nozzle; contain 40% of the initial annihilation energy
- Very high Isp (~ 28 million seconds) but low thrust (typically 10's N)
- Typical $\bar{p}$ flow rate ~ 100's µg/sec

Figure: Forward, R. L., *Antiproton Annihilation Propulsion,* AFRPL TR-86-034, AFRPL/LKC, Edwards AFB, Ca., Sep. 1985.